Mandala Meditation Book For Adults
Large Print Deep Coloring Book
For Mindfullness, Relaxation, and Stress Relief

By Peaceful Mind Adult Coloring Books

Copyright © 2020

All rights reserved. No part of this publication may be reproduced, distributed, or transmitted in any form or by any means, including photocopying, recording, or other electronic or mechanical methods, without the prior written permission of the publisher

This book belongs to:

_____

Calming Mandalas to Color

A mandala is a powerful symbol that is used in meditation, sacred art, and ceremonial rituals. The term mandala literally translates to "circle" in Sanskrit, but the name derives from two different words: "la" (container) and "manda" (essence). Mandalas are used by Hindus and Buddhists to represent the universe. These beautiful circles are a representation of life and our relation to the infinite that extends beyond and within ourselves.

The mandala designs in this book can be used for stress relief, relaxation, meditation, calming color therapy or just for fun! These mandalas are meant to be simple and easy for beginners, but also great for experienced artists who delight in the beauty of these harmonious circle designs.

There are 30 meditative and calming mandala images in this book that you can use as color therapy, plus 4 bonus mandalas from other Color Questopia mandala books. Let go of all anxiety and worries as you indulge your creativity. Bring peace to your soul.

You can color these meditative patterns using colored pencils, markers, watercolors, gel pens, paint, or a box of crayons. However, if you use markers or pens, make sure to place a blank sheet of paper under the coloring page so that the ink doesn't bleed through.

Let go of all stress and relax your mind. Simply let the mandalas facilitate healing, mindfulness, balance and tranquility within you.

1.

3.

4.

5.

6.

7.

8.

10.

13.

14.

16.

17.

18.

20.

22.

23.

24.

25.

26.

28.

29.

ENJOY BONUS IMAGES FROM SOME OF OUR OTHER RELAXING MANDALA BOOKS!

FIND ALL OF OUR BOOKS ON AMAZON

Mandala Pattern Book - SImple Designs For Adults Easy Adult Coloring Mandala Designs

5.

16.

Mandala Coloring Book For Adults Relaxation and Stress Relief

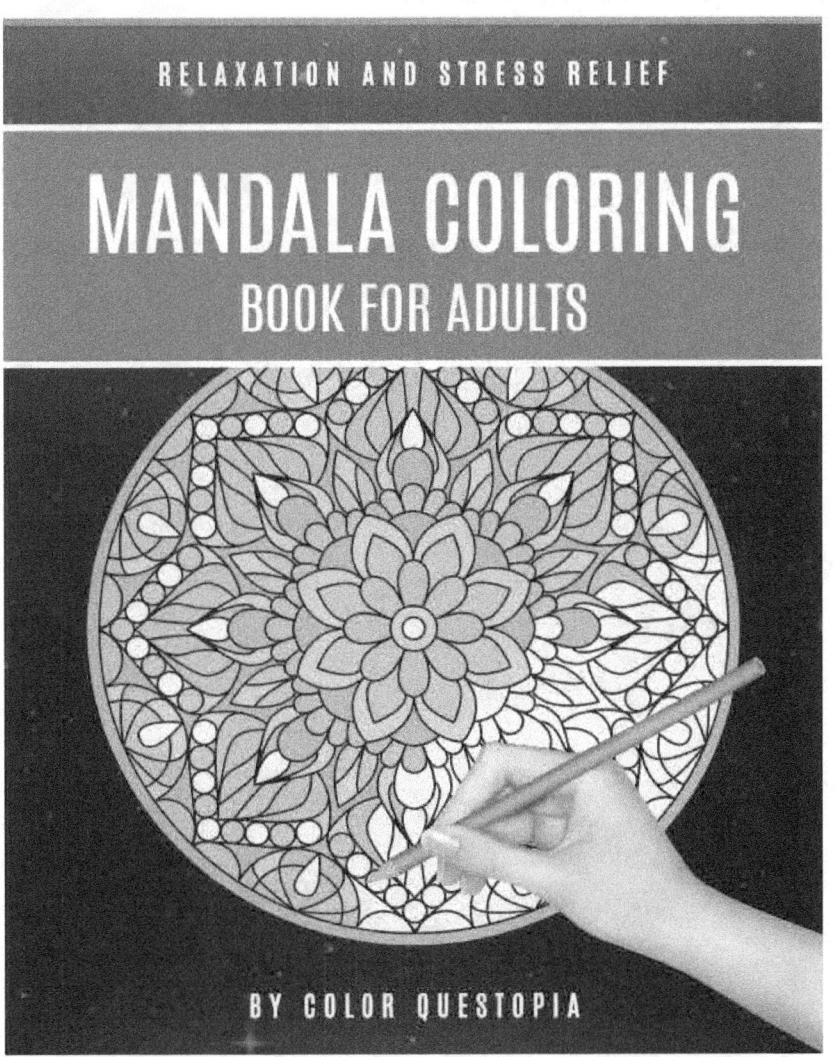

29.

19.

www.ingramcontent.com/pod-product-compliance
Lightning Source LLC
Chambersburg PA
CBHW080519220526
45465CB00006B/2535